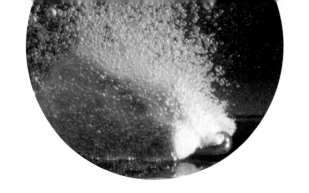

CORE CHEMISTRY

Chemical Reactions

DENISE WALKER

A⁺
Smart Apple Media
an imprint of Black Rabbit Books

This book has been published in cooperation with Evans Publishing Group.

Series editor: Harriet Brown, Editor: Harriet Brown, Design: Simon Morse, Illustrations: Ian Thompson, Simon Morse

Published in the United States by Smart Apple Media
2140 Howard Drive West
North Mankato, Minnesota 56003

Library of Congress Cataloging-in-Publication Data

Walker, Denise.
Chemical reactions / by Denise Walker.
p. cm. – (Core chemistry)
Includes index.
ISBN 978-1-58340-820-9
1. Chemical reactions. I. Title.

QD501.W244 2007
541'.39—dc22   2007005178

9 8 7 6 5 4 3 2 1

Contents

Introduction

Chemical reactions are continually taking place on our planet. They occur inside our bodies, inside plants and animals, in the atmosphere, and beneath the earth's surface. Scientists know a great deal about chemical reactions and have discovered how to adapt them for our own use.

This book takes you on a journey to discover more about chemical reactions; find out how they work, what they can do, and why we need them. Learn all about reactions that give off heat and find out why some reactions freeze. Take a closer look at chemical equations and see that they are not as complicated as they might appear. You can also learn the difference between physical and chemical changes, discover how to start a chemical reaction, and how to speed it up.

This book also contains feature boxes that will help you unravel more about the mysteries of chemical reactions. Test yourself on what you have learned so far; investigate some of the concepts discussed; learn key facts; and discover some of the scientific findings of the past and how these might be utilized in the future.

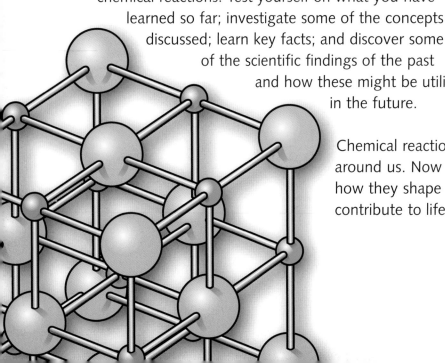

Chemical reactions are all around us. Now you can understand how they shape our planet and contribute to life as we know it.

DID YOU KNOW?

▶ Look for these boxes. They contain interesting facts about chemical reactions in the world around us.

TEST YOURSELF

▶ Use these boxes to see how much you've learned. Try to answer the questions without looking at the book, but take a look if you are really stuck.

INVESTIGATE

▶ These boxes contain experiments that you can carry out at home. The equipment you need is usually inexpensive and easy to find around the house.

TIME TRAVEL

▶ These boxes describe scientific discoveries from the past and fascinating developments that pave the way for the advance of science in the future.

ANSWERS

On pages 46 and 47, you will find the answers to the questions from the "Test yourself" and "Investigate" boxes.

GLOSSARY

Words highlighted in **bold** are described in detail in the glossary on pages 46 and 47.

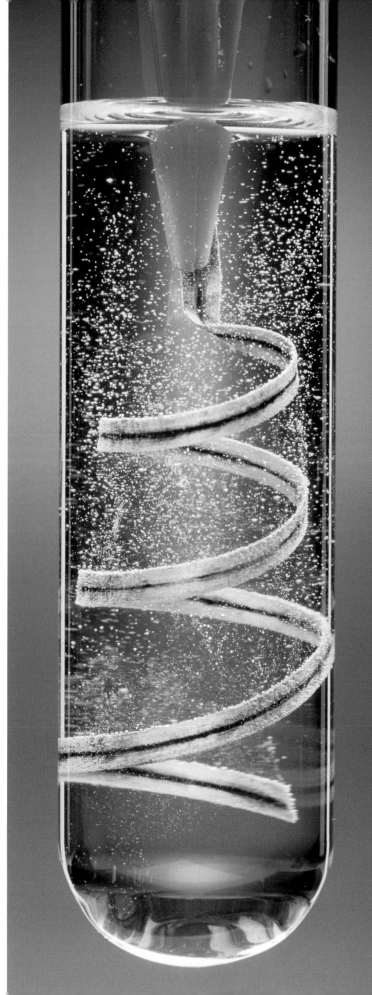

1

Physical and chemical changes

Many substances change naturally over time. Rocks change gradually as a result of weathering, and water changes when its temperature rises or falls. Some substances change when other substances join them as a result of chemical reactions. Any change can be classified as either a physical or a chemical change.

PHYSICAL CHANGE

Physical change occurs when the properties of a substance change but its chemical composition does not. The properties of a substance are how it looks, feels, or behaves. Consider an ice cube that **melts** when it is taken out of the freezer. Ice is frozen water and as it melts, it changes into liquid water. Its physical state has changed, but it is still water.

CHANGING FROM SOLID TO LIQUID TO GAS

Solid

Liquid

Gas

A water molecule consists of two hydrogen atoms joined together with one oxygen atom: H_2O. When ice melts, the forces holding the molecules together weaken but the molecules themselves remain unchanged. This is a physical change. If liquid water is heated, it will **boil** and turn to gaseous water, called steam. During boiling, the molecules gain energy from heat. The weak bonds between the water molecules in the liquid break, and they are able to move away from each other. But they are still water molecules. Again, this is a physical change.

Physical changes can be easily reversed as we see during **condensation** and **freezing**. When molecules cool, they lose energy and become less active. This means that they do not move as far from each other, and they can bond again.

◄ The particles in a solid are tightly bound together. The particles in a liquid are only loosely bound, and in a gas, they are not bound at all.

MISCIBLE AND IMMISCIBLE

Combining chemicals can also result in a physical change, but only if the chemical composition does not change. When oil and water are combined there is no chemical change. Oil and water do not mix—they are **immiscible**. The forces between the water molecules are much stronger than those between the water and the oil molecules, so the substances do not mix.

▲ Oil floats on water. The two substances do not mix.

Liquids that do mix, such as ink and water, are called **miscible** substances. The forces between the ink and the water molecules are the same, so they mix. However, they are still undergoing only a physical change. The ink molecules and the water molecules do not change their chemical form.

Physical changes include the alteration of:

▶ State of matter
▶ Appearance
▶ Strength
▶ Hardness
▶ How well the substance conducts electricity
▶ Size

CHEMICAL CHANGE

A chemical change occurs when the chemical composition of a substance changes. This is usually caused by a chemical reaction. Chemical reactions begin with one or more substances, called **reactants**, and end with at least one completely new substance, or **product**.

Chemical reactions can be sudden, or they can take place over long periods of time. They may be obvious or unnoticeable. To determine whether a chemical change has occurred, scientists study the chemicals present before and after the reaction.

CHEMICAL BONDS

During a chemical change, **chemical bonds** break and form. For example, when salt forms, sodium atoms react with chlorine molecules. This chemical reaction produces sodium chloride, which is the salt that we put on our food. Chemical bonds have formed between the sodium and the chlorine.

Chemical changes are more permanent than physical changes. They cannot be easily reversed because chemical bonds are not easy to break. An example of this is the formation of rust. The chemical name for rust is hydrated iron (III) oxide. It forms through a reaction between iron, water, and oxygen. It would seem an easy solution to just scrape the rust from our bicycles to reveal the shiny, strong metal beneath. However, the metal beneath will be thin and **corroded**. Rusting is not easily reversed, and further rust can only be prevented by coating the metal so that oxygen and water cannot attack it.

EXPLOSIONS

Chemical reactions can be extremely dramatic, noisy, and colorful. An explosion is a chemical reaction that involves a high temperature, the sudden and violent release of energy and light, and the production of gases. Explosions are **combustion** (burning) reactions and require the presence of oxygen. They are often associated with death and destruction, but this is not their only purpose. Explosions are essential to the mining industry to blast apart rocks in search of valuable minerals from the earth's crust. Explosions also provide thrust in rockets and jet engines, and they are the basis for fireworks.

Traditional fireworks, such as the firecrackers used for Chinese New Year celebrations, contain gunpowder. Gunpowder is 75 percent potassium nitrate, 15 percent charcoal, and 10 percent sulfur. When these chemicals are ignited, they undergo the following chemical reactions:

▶ Charcoal burns to produce a mixture of carbon dioxide and carbon monoxide.

▶ Potassium nitrate releases oxygen.

▶ Sulfur reacts with oxygen to produce sulfur dioxide.

Because the chemicals are contained in a paper tube, the pressure of the gaseous products quickly builds, and the firecracker "explodes." The result is a very loud noise and a lot of smoke.

▼ This firecracker is exploding. The chemicals are enclosed in cylindrical containers. They are connected by a fuse along which a spark travels.

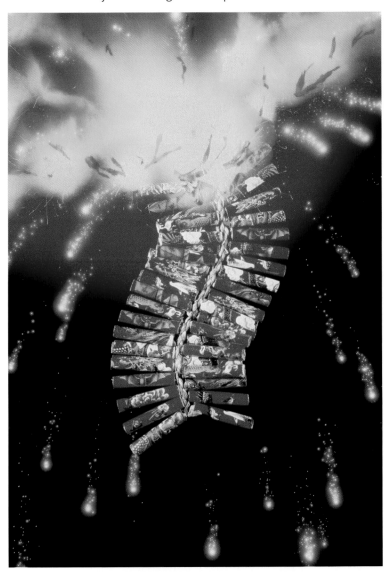

EXPLOSIONS IN CARS

When a gasoline engine is started, the gasoline and the oxygen in the air combine, and they are ignited by the spark plugs. The result of this chemical reaction is a small explosion that releases carbon dioxide, nitrogen oxides, and steam. These gaseous products force a piston to move, which turns parts of the engine, and then the wheels, causing the car to move.

EXPLOSIONS IN ROCKETS

Rockets are also propelled using explosions, but instead of producing a turning action, the products of the explosion are directed one way, forcing movement in the opposite direction. This is a little like running water through a hose without holding it at the end. The force of the water's flow out of the hose causes the hose to twist and turn wildly.

Rockets use solid or liquid fuel. The Chinese developed the earliest rockets for warfare in the 1200s. These rockets were ignited with gunpowder. Today, liquid fuels power the main engines of space shuttles, and spacecraft such as Cassini. In 2004, Cassini arrived on Saturn to study its moons.

Liquid fuels include a liquid hydrogen and liquid oxygen mixture and a kerosene and liquid oxygen mixture. The fuel and an **oxidizer** are pumped into a chamber where they are ignited. When they burn, the fuel and oxidizer react to create a high-pressure, high-speed plume of gases. Pumps accelerate the gases until they are speeding up to 9,942 miles (16,000 km) per hour. The force of the gases leaving the rocket propels it forward.

▼ This rocket burns solid and liquid fuel to lift the shuttle into orbit, 398 miles (640 km) above the earth's surface.

INVESTIGATE

Determine whether the following changes are physical or chemical.
(1) Stage smoke—solid carbon dioxide turns to gaseous carbon dioxide.
(2) Hydrogen and oxygen changes into water.
(3) Hand warmers.

How to start a chemical reaction

The atoms within a molecule are held together by chemical bonds. In order for reactants to become products, the chemical bonds must be broken. This requires energy. To begin a chemical reaction, energy is added in the form of heat, motion, water, or light.

STARTING A REACTION

(1) HEAT
Heat energy causes the reactants' atoms to move. As they move, the bonds between the atoms begin to break. When heat is added, the molecules collide with one another more frequently. This also breaks the chemical bonds.

(2) MOTION
Stirring can provide enough energy to trigger the bonds to break. Some chemicals are sensitive to touch and will react to the slightest motion.

(3) WATER
Some metals are very reactive when exposed to water. Potassium is extremely reactive with water and the reaction releases a lot of heat energy. This heat accelerates the reaction. In fact, the strong reaction of potassium with water is an extremely fast and violent version of rusting.

(4) LIGHT
Light energy can break chemical bonds and start a reaction. Nondigital camera film is a paper soaked in light-sensitive silver chloride. When a photo is taken, light enters the camera and darkens the silver chloride. This leaves an image called a negative. The negative can be converted into a photograph.

Light initiates the most important chemical reaction on our planet—photosynthesis. This is a reaction in plants that occurs between carbon dioxide and water.

◀ The reaction between potassium metal and water produces enough heat to melt the potassium.

WHAT HAPPENS DURING A CHEMICAL REACTION?

The input of energy frees atoms from their compounds. They look for new atoms to bond with. The free atoms force the other reactants to release their chemical bonds. The released atoms rearrange themselves into new chemical combinations that we call products.

ENERGY

During a chemical reaction, energy changes its form. Initially, chemical energy is stored in the chemical bonds. The reaction begins once heat energy, light energy, or motion energy break some of the bonds. This releases chemical energy. Energy is produced in the form of sound or light, and heat energy is lost or gained. At the end of the reaction, energy is again contained within chemical bonds.

There is usually less energy within the chemical bonds of the products than there is in the chemical bonds of the reactants. Most reactions would not take place if this was not the case. For example, a ball will roll down a hill because it has more energy at the top of the hill than at the bottom. It will not spontaneously roll up a hill because it does not have enough energy. The same is true for chemical reactions.

TEST YOURSELF

▶ What starts each of the following reactions?

(1) When exposed to a photographic flash, a mixture of chlorine and hydrogen react to produce hydrogen chloride.

(2) Iron and oxygen will react to form rust. This is a very slow reaction.

When dealing with chemical reactions, it is important to know the following terms:

Type of particle	Definition
Atom	Simplest form of a substance
Molecule	Two or more atoms chemically bonded
Ion	An atom with a positive or negative charge, or a small group of charged atoms
Element	A substance made from one type of atom that cannot be broken down chemically
Compound	Two or more elements chemically bonded

TIME TRAVEL: DISCOVERIES OF THE PAST

In 1826, a British chemist named John Walker observed that motion could cause a chemical reaction. He stirred potassium carbonate and antimony (a metal) together, and then scraped the stirring stick on the floor to remove some of the mixture. To his surprise, the stick caught fire. This prompted the invention of matches. Antimony was replaced with phosphorus. Unfortunately, the first matches were dangerous because the phosphorus poisoned the matchmakers. The matchmakers' clothes also caught fire because of the volatile nature of the chemical mixture. Safety matches were invented in 1844. The match carries a reactant—potassium chlorate—and the strip on the box contains phosphorus. The match ignites when it is scraped on the phosphorus strip.

Exothermic and endothermic reactions

During a chemical reaction, energy changes from one form to another. Chemical bonds break and form again. We cannot see these changes happen, but we can observe clues such as color changes, sounds, and smells. One of the most common changes is a change in temperature. If we place a sensitive thermometer in a chemical reaction, we can see that it has either warmed, an **exothermic** reaction, or cooled, an **endothermic** reaction. Thermodynamics is the study of the conversions between heat and other forms of energy.

EXOTHERMIC REACTIONS

Reactions that release heat energy are called exothermic reactions. These reactions feel warm to the touch. If a reaction is highly exothermic, it may be dangerous to touch. Explosions are extreme exothermic reactions. We use a thermometer or temperature probe to detect temperature changes.

Energy is required to break chemical bonds, which means that energy is spent, or used. As the atoms rearrange themselves and form new chemical bonds, energy is released, often in the form of heat. In an exothermic reaction, the total energy released in the formation of new bonds is greater than the total energy used to break the bonds. Therefore, the reaction causes an increase in temperature.

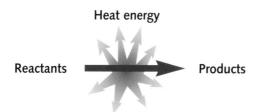

Exothermic reactions are common because the products have less energy stored in the chemical bonds than the reactants. This encourages exothermic reactions because one of the laws of thermodynamics states that all reactants and products will try to contain as little energy as possible.

Combustion reactions are exothermic (see pages 22–23). When methane burns in oxygen, it produces carbon dioxide and water.

Methane contains more energy than carbon dioxide and water. This extra energy is released as heat when methane breaks down.

▲ Landfills release methane gas. The gas can build up and explode in a violent exothermic reaction. To prevent this, the methane is slowly burned.

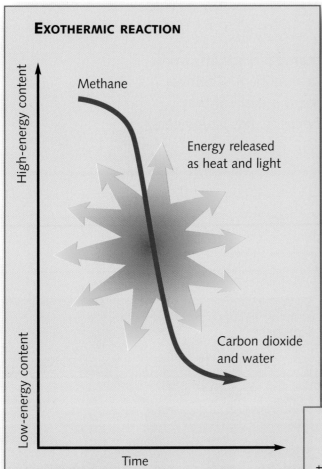

EXOTHERMIC REACTION

High-energy content

Methane

Energy released as heat and light

Low-energy content

Carbon dioxide and water

Time

▲ This graph shows that there is more energy in the reactants than in the products. Energy is released during an exothermic reaction.

ENDOTHERMIC REACTIONS

Chemical reactions that require heat energy are called endothermic reactions. These reactions become cooler as they progress. Energy is added to the reactant mixture to break the initial molecular bonds, and energy is released as new chemical bonds form. Overall, more energy is absorbed by the reaction than is released. The products of endothermic reactions contain more energy than the reactants.

When barium hydroxide is mixed with ammonium chloride, the reaction is so endothermic that it can freeze water.

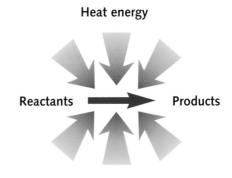

Heat energy

Reactants ➝ Products

Another endothermic reaction occurs when an egg is cooked. The egg absorbs heat energy that is supplied by a stove. If the heat energy is not added, the reaction will not take place.

▼ This graph shows that there is less energy in the reactants than in the products. Energy is absorbed during an endothermic reaction.

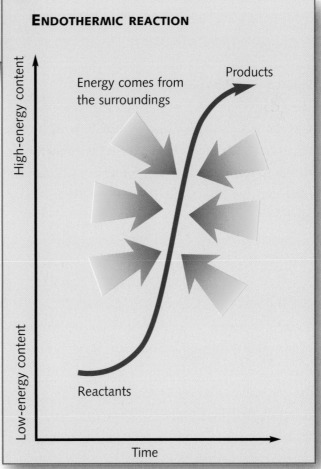

ENDOTHERMIC REACTION

High-energy content

Energy comes from the surroundings

Products

Low-energy content

Reactants

Time

4

Understanding equations

Although there are many words to describe chemical reactions, words are not always the most efficient way of explaining them. Chemists around the world need to share their ideas, and it is easier if they speak the same "language." In the case of chemistry, this language is chemical formulas and chemical equations.

WORD EQUATIONS

In a word equation, the names of the reactants are written on the left and the names of the products on the right. The reactants and products in the equation are separated by the symbol ⟶ which means "goes to." In other words, in a chemical reaction, the reactants "go to" the products.

IRON AND SULFUR

Iron and sulfur are both elements. When they react they produce a compound called iron sulfide. The word equation for this reaction is:

Iron + Sulfur ⟶ Iron sulfide

▲ Iron (left) and sulfur (right).

Iron and sulfur will only react to form iron sulfide if heat is added. The heat is responsible for breaking the bonds present within each of the elements. The atoms that have been liberated react with each other. Heat is neither a reactant nor a product. Therefore, we write the equation like this:

Iron + Sulfur $\xrightarrow{\text{heat}}$ Iron sulfide

▼ Iron sulfide is a hard and nonmagnetic compound.

Iron and sulfur are elements with their own distinct properties. Sulfur is a yellow powder with a strong smell. Iron is a shiny magnetic metal. When they are combined, their properties do not exist in the iron sulfide. If a magnet is held close to the compound, nothing happens. The product has different properties from the reactants. A chemical change has occurred.

HYDROGEN AND CHLORINE

Hydrogen and chlorine gas react slowly, but when they are exposed to ultraviolet (UV) light, the reaction can become explosive. This reaction is represented as follows:

$$\text{Hydrogen} + \text{Chlorine} \xrightarrow{\text{UV}} \text{Hydrogen chloride}$$

HYDROGEN PEROXIDE

Hydrogen peroxide is a colorless liquid that decomposes to produce water and oxygen. The process is slow but can be accelerated with a compound called manganese dioxide. This chemical is a **catalyst**. A catalyst is a substance that alters a chemical reaction (usually by speeding it up) without being changed itself. Manganese dioxide is not used up, so it is neither a reactant nor a product. In fact, at the end of the reaction, the black powder of manganese dioxide can be seen at the bottom of the test tube. The word equation for this change is:

$$\text{Hydrogen peroxide} \xrightarrow{\text{Manganese dioxide}} \text{Water} + \text{Oxygen}$$

▶ This flask contains hydrogen peroxide and some raw liver. Liver contains an **enzyme** (a biological catalyst) called catalase. It accelerates the breakdown of hydrogen peroxide in a way similar to manganese dioxide. You can see the bubbles of oxygen that have formed.

TEST YOURSELF

▶ Write word equations to represent the following chemical changes.
Try to include as much detail as you can.

(1) Carbon reacts with oxygen in the presence of heat to form carbon dioxide.
(2) Magnesium burns in oxygen to form magnesium oxide.
(3) Methane and chlorine react in the presence of ultraviolet light to form chloromethane and hydrogen chloride.
(4) Hydrogen peroxide decomposes to water and oxygen in the presence of the catalyst catalase.

CHEMICAL FORMULAS

Word equations are a useful way to represent a chemical change. However, word equations are also too simple. They do not tell us exactly how the atoms rearrange themselves. For a more complete understanding of a chemical change we must write a chemical equation. Chemical equations contain chemical formulas, rather than words.

ELEMENTS IN CHEMICAL REACTIONS

If elements are present as either reactants or products, it is more efficient to use their chemical symbols. The chemical symbols for the elements are listed on the periodic table.

Some elements exist as pairs of atoms. When they are present in a chemical equation, we write them with a small "2" following their symbol, as shown in the table above right.

Element	Written in an equation as:
Chlorine	Cl_2
Oxygen	O_2
Nitrogen	N_2
Hydrogen	H_2
Bromine	Br_2
Fluorine	F_2
Iodine	I_2

ELEMENTS AND THEIR SYMBOLS

Here is a list of the more common elements.

Ac = Actinium	Ga = Gallium	Pb = Lead
Ag = Silver	Ge = Germanium	Pd = Palladium
Al = Aluminum	H = Hydrogen	Pt = Platinum
Ar = Argon	He = Helium	Ra = Radium
As = Arsenic	Hf = Hafnium	Rb = Rubidium
At = Astatine	Hg = Mercury	Rh = Rhodium
Au = Gold	I = Iodine	Rn = Radon
B = Boron	Ir = Iridium	S = Sulfur
Ba = Barium	K = Potassium	Sb = Antimony
Be = Beryllium	Kr = Krypton	Sc = Scandium
Br = Bromine	Li = Lithium	Se = Selenium
C = Carbon	Mg = Magnesium	Si = Silicon
Ca = Calcium	Mn = Manganese	Sn = Tin
Cd = Cadmium	Mo = Molybdenum	Sr = Strontium
Cl = Chlorine	N = Nitrogen	Ta = Tantalum
Co = Cobalt	Na = Sodium	Ti = Titanium
Cr = Chromium	Nb = Niobium	V = Vanadium
Cs = Cesium	Ne = Neon	W = Tungsten
Cu = Copper	Ni = Nickel	Xe = Xenon
F = Fluorine	O = Oxygen	Zn = Zinc
Fe = Iron	P = Phosphorus	Zr = Zirconium

THE PERIODIC TABLE

1 H	

□ Metals
□ Nonmetals

I   II —— Groups

Atomic number

| 2
He |

| 3
Li | 4
Be |
| 11
Na | 12
Mg |

| 5
B | 6
C | 7
N | 8
O | 9
F | 10
Ne |
| 13
Al | 14
Si | 15
P | 16
S | 17
Cl | 18
Ar |

19 K	20 Ca	21 Sc	22 Ti	23 V	24 Cr	25 Mn	26 Fe	27 Co	28 Ni	29 Cu	30 Zn	31 Ga	32 Ge	33 As	34 Se	35 Br	36 Kr
37 Rb	38 Sr	39 Y	40 Zr	41 Nb	42 Mo	43 Tc	44 Ru	45 Rh	46 Pd	47 Ag	48 Cd	49 In	50 Sn	51 Sb	52 Te	53 I	54 Xe
55 Cs	56 Ba	57 La	72 Hf	73 Ta	74 W	75 Re	76 Os	77 Ir	78 Pt	79 Au	80 Hg	81 Tl	82 Pb	83 Bi	84 Po	85 At	86 Rn
87 Fr	88 Ra	89 Ac	104 Rf	105 Db	106 Sg	107 Bh	108 Hs	109 Mt	110 Ds	111 Rg

| 58
Ce | 59
Pr | 60
Nd | 61
Pm | 62
Sm | 63
Eu | 64
Gd | 65
Tb | 66
Dy | 67
Ho | 68
Er | 69
Tm | 70
Yb | 71
Lu |
| 90
Th | 91
Pa | 92
U | 93
Np | 94
Pu | 95
Am | 96
Cm | 97
Bk | 98
Cf | 99
Es | 100
Fm | 101
Md | 102
No | 103
Lr |

COMPOUNDS IN CHEMICAL EQUATIONS

Some simple compounds can be formulated using a method called "swap and drop."

The components of a certain type of compound are called ions. Ions form when atoms gain or lose negatively charged electrons during chemical reactions. For example, sodium atoms each lose an electron to become sodium ions. Sodium atoms have the symbol Na. Therefore, sodium ions have the symbol Na^+. By losing one electron, the sodium atom gains a positive charge because it now has one more proton than electrons. Remember, protons have a positive charge and electrons have a negative charge.

Sodium atoms will always lose only one electron. This is because they have one electron in the outer shell of their atoms. Moving toward the nucleus of the atom, the second shell has a complete set of eight electrons. If the sodium atom loses its outermost electron, the second shell becomes the outer shell. Atoms with an outer shell full of electrons are much more stable than those with partially filled shells.

SWAP AND DROP:

SODIUM + CHLORINE

When sodium and chlorine react, each sodium atom loses an electron to a chlorine atom. This forms sodium ions, Na^+, and chloride ions, Cl^-. Use the following steps to determine the formula of the compound.

(1) Write the number of the charge beneath the ion. The symbols $^+$ and $^-$ refer to $^{1+}$ and $^{1-}$.

(2) Swap the numbers around and "drop" them to the ground.

(3) This tells us that we need one sodium ion to balance one chloride ion. So the formula is NaCl.

MAGNESIUM + CHLORINE

To gain a full outer shell of electrons, magnesium must lose two electrons. As an ion it has two more protons than electrons. Its formula is Mg^{2+}.

Using swap and drop, one Mg ion is needed to balance two Cl ions. The formula is $MgCl_2$.

Here are the symbols and charges for some common ions:

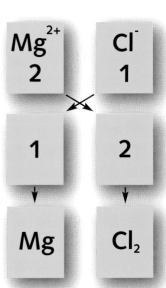

3+	2+	1+	1-	2-
Aluminum Al^{3+}	Magnesium Mg^{2+}	Sodium Na^+	Chloride Cl^-	Oxide O^{2-}
Iron(III) Fe^{3+}	Iron Fe^{2+}	Lithium Li^+	Bromide Br^-	Sulfide S^{2-}
Manganese(III) Mn^{3+}	Copper Cu^{2+}	Potassium K^+	Hydroxide OH^-	Carbonate $CO_3{}^{2-}$
Chromium(III) Cr^{3+}	Calcium Ca^{2+}	Hydrogen H^+	Nitrate $NO_3{}^-$	Sulfate $SO_4{}^{2-}$

The formulas for the compounds in the table below cannot be easily determined. Use this table as a reference:

Name	Formula
Water	H_2O
Carbon dioxide	CO_2
Hydrochloric acid	HCl
Sulfuric acid	H_2SO_4
Sodium hydroxide	$NaOH$

TEST YOURSELF

▶ Use the "swap and drop" method to write the formulas for the following compounds.

(1) Copper oxide

(2) Iron chloride

(3) Aluminum bromide

(4) Calcium oxide

BALANCING CHEMICAL EQUATIONS

In every chemical reaction, reactants turn into products. No atoms are lost or gained during a chemical reaction. Every atom should be accounted for on both sides of the equation; all chemical equations are balanced. For example, the following chemical change is a common burning reaction.

Carbon + Oxygen ⟶ Carbon dioxide
$$C + O_2 \longrightarrow CO_2$$

One atom of carbon combines with two atoms of oxygen to produce one molecule of carbon dioxide.

On each side of the equation there is one carbon atom and two oxygen atoms. The equation is balanced.

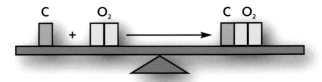

Use the "swap and drop" method from the previous page and the table of common formulas to work through the following steps for balancing equations.

(1) Write a word equation.

Potassium + Water ⟶ Potassium hydroxide + Hydrogen

(2) Convert the word equation into symbols.

$$K + H_2O \longrightarrow KOH + H_2$$

(3) Count the number of each type of atom on each side of the arrow in the equation. A small "2" next to the H means there are two atoms of hydrogen. In this example we have:

On the left side:	On the right side:
1 x K	1 x K
1 x O	1 x O
2 x H	3 x H

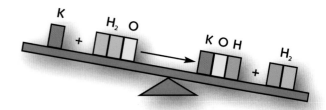

If the numbers on each side are the same, no adjustments are necessary. If not, balance the equation by using a coefficient at the beginning of a formula. The coefficient means that all atoms in the formula are multiplied by it. In this example, the number of hydrogen atoms is unbalanced. Balance the equation by inserting the following coefficients.

$$2K + 2H_2O \longrightarrow 2KOH + H_2$$

We now have the following:

On the left side:	On the right side:
2 x K	2 x K
2 x O	2 x O
4 x H	4 x H

The number of atoms on each side is the same and the equation is balanced.

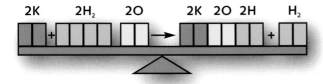

STATE SYMBOLS

These are symbols that show the physical state of each of the reactants and products:

State symbol	What does it mean?
(s)	Solid
(l)	Liquid
(g)	Gas
(aq)	Aqueous (dissolved in water)

By adding the state symbols, our example now becomes:

$$2K_{(s)} + 2H_2O_{(l)} \longrightarrow 2KOH_{(aq)} + H_{2(g)}$$

To determine the natural state of the reactants and products, look them up in a reference book or on the Internet.

The final equation shows that:
(1) Two atoms of potassium react with two molecules of water.
(2) The products are two molecules of potassium hydroxide and one molecule of hydrogen.
(3) The potassium is solid, the water is liquid, and the potassium hydroxide is aqueous. Bubbles of hydrogen gas may be present.

TEST YOURSELF

▶ Write balanced chemical equations for each of the following reactions:

(1) Hydrogen reacts with oxygen to form water.

(2) Sodium reacts with chlorine to form sodium chloride.

(3) Magnesium reacts with oxygen to form magnesium oxide.

▼ This magnesium ribbon is burning in air, which contains oxygen, to form magnesium oxide.

5

Decomposition, precipitation, and combustion

Millions of chemical reactions occur every day. Reactions in our bodies keep us alive and healthy, while reactions in our vehicles make them move. Chemists classify reactions according to how they work. Three common types of chemical reactions are thermal decomposition, precipitation, and combustion.

▲ These chalk cliffs are mainly made from calcium carbonate, or limestone.

THERMAL DECOMPOSITION

Decomposition is when a single substance breaks down into two or more products. If this change occurs under the influence of heat, then it is called thermal decomposition. Decomposition reactions can also be caused by light.

Calcium carbonate (limestone) undergoes thermal decomposition when it is heated. This reaction produces calcium oxide and carbon dioxide. Calcium oxide is also known as lime. Lime is used in agriculture to neutralize acidic soil.

HYDRATED COMPOUNDS

Hydrated compounds undergo thermal decomposition. They have water molecules in their structures. When hydrated copper sulfate is heated, the water is driven out. It has the formula $CuSO_4 \cdot 5H_2O$ and is a crystal. When it is heated, the water **evaporates** and leaves behind anhydrous—without water—copper sulfate. This thermal decomposition is characterized by a color change from blue to grey/white.

◄ Hydrated copper sulfate is blue.

▼ Anhydrous copper sulfate is grayish white.

Anhydrous copper sulfate creates dry environments by absorbing water from the atmosphere. It is similar to the packets of silica gel that are found in a new pair of shoes. Silica gel keeps the shoes dry and prevents the leather from becoming damaged.

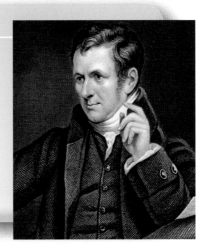
PRECIPITATION REACTIONS

When two solutions are mixed, they react to form an insoluble product called a **precipitate**. It can appear as a fine powder or a dense, gel-like mass. If the precipitate is fine and floats in liquid, it is called a **suspension**.

Precipitation reactions identify components of unknown compounds. For example, precipitation tests for the presence of metals. When sodium hydroxide solution is added to certain metal compounds, a distinct, colored precipitate is observed as outlined in the table below.

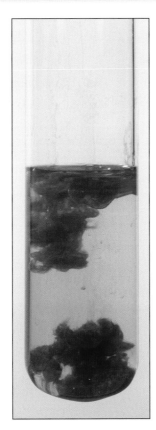

◀ This image shows a dense, gel-like precipitate of iron (III) hydroxide.

Metal	Color of precipitate	Name of precipitate
Copper	Blue	Copper hydroxide
Aluminum	White	Aluminum hydroxide
Iron	Rusty brown	Iron (III) hydroxide
Iron	Green	Iron (II) hydroxide

COMBUSTION REACTIONS

A combustion reaction results in a chemical change. It usually involves a reaction between a substance and oxygen from the air and releases heat, light, and other chemical products. Combustion reactions are very important. We combust, or burn, fuels to power our vehicles.

INGREDIENTS OF A COMBUSTION REACTION

All combustion reactions require three things: oxygen, fuel, and heat. These are represented in the "fire triangle." If one of these ingredients is removed, then the fire will go out. Knowing this helps chemists create fire extinguishing solutions.

FIRE TRIANGLE

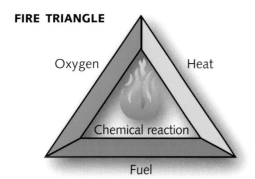

(1) HEAT

Many combustion reactions require an ignition source. Campfires and grills are lit with matches or an electronic ignition system. To start a fire, the heat must be sufficient to break the initial chemical bonds and allow the atoms to begin rearranging themselves. As the atoms change into products, more energy is released than was introduced. All combustion reactions are exothermic.

(2) FUEL

When fuel ignites, it reacts with oxygen. Once the fuel is exhausted, a fire will go out. When forest fires burn, the trees provide a large supply of fuel. Forest fires can rage for weeks at a time, causing much destruction. Every year in the United States, 6,718 square miles (17,400 sq km) of land are burned in forest fires. This is the equivalent of around 0.18 percent of land in the U.S., or approximately 2.5 million soccer fields.

▼ Forest fires are powerful exothermic reactions. They produce intense heat and lots of smoke.

(3) Oxygen

Air contains the oxygen necessary for a fire. Around 20 percent of the air is oxygen gas. Some flammable chemicals are stored in an oxygen-free atmosphere to ensure the chemicals do not ignite. Some metals, such as sodium and potassium, are so reactive with oxygen that they are stored in oil; they cannot be stored in water because its oxygen content would be enough to ignite the metals.

If there is sufficient oxygen, complete combustion occurs. This means that the maximum amount of chemical product is produced. When there is an insufficient amount of oxygen present, incomplete combustion occurs. This happens in the exhausts of old or poorly functioning cars.

Gasoline contains the element carbon. When carbon is burned in sufficient oxygen, it produces carbon dioxide gas. However, if it undergoes incomplete combustion, it will also release carbon monoxide gas and small amounts of black carbon deposits called **particulates**. Although fatal at high levels, carbon monoxide from car exhausts escapes into the atmosphere where it quickly disperses by **diffusion**.

Putting out a fire

Some combustion reactions can rage out of control. To manage a fire, the basic rules of the fire triangle must be followed. Removing one side of the triangle will put out the fire. How effective this is depends on how long the fire has been burning and how much fuel is present.

It is important to know how fires can be safely extinguished. Fires that involve burning oil or gasoline must be extinguished with sand, soil, or carbon dioxide. These deprive the fire of oxygen, and it will quickly go out. If water is used, the fuel will float on the water and cause the fire to spread.

Fires caused by sparks from electrical appliances should not be extinguished using water or foam. Both of these substances conduct electricity and could cause a serious electrical shock. Carbon dioxide and dry chemical extinguishers should be used to put out electrical fires. Before fighting an electrical fire, disconnect the electricity supply.

Test yourself

▶ Think of four ways you rely on combustion reactions in your everyday life, such as when you are traveling, at home, or at school.

▶ Incomplete combustion is occurring in this bus. The exhaust fumes contain carbon monoxide gas and black carbon particulates.

Displacement reactions

In **displacement** reactions, the chemicals swap their chemical partners. For example, in the following reaction, potassium bromide reacts with chlorine. This produces potassium chloride (a compound) and bromine (an element).

Potassium bromide + Chlorine ⟶ Potassium chloride + Bromine

The chlorine has displaced the bromine from its compound and bonded with the potassium.

OXIDATION AND REDUCTION

Displacement reactions are composed of two parts: an oxidation reaction and a reduction reaction. In the example above, the chlorine has been reduced and the bromine has been oxidized. When chemical reactants reorganize themselves into products, the new bonds form as a result of the transfer of electrons between atoms. This rearrangement of electrons is called bonding. Electrons cannot be lost from one atom unless there is another nearby to bond with them.

Oxidation usually involves the loss of an electron, and reduction involves the gain of an electron. Reduction and oxidation always occur together. This is sometimes simplified to the term **redox**. All displacement reactions involve an oxidation component and a reduction component, so they are called redox reactions.

▶ This is an iron ore quarry. Redox reactions are used to extract iron from its ore. The ore is put in a blast furnace and the following reaction takes place:

Iron oxide + Carbon monoxide ⟶ Iron + Carbon dioxide

UNDERSTANDING OXIDATION AND REDUCTION

OXIDATION – occurs when any of the following happens:

▶ Electrons are lost
▶ Oxygen is gained
▶ Hydrogen is lost

Only one of these criteria needs to be met.

REDUCTION – occurs when any of the following happens:

▶ Electrons are gained
▶ Oxygen is lost
▶ Hydrogen is gained

The movement of electrons is easily remembered using **OIL RIG**. This is an acronym for Oxidation **I**s **L**oss of electrons and Reduction **I**s **G**ain of electrons.

The example of a displacement reaction at the beginning of this chapter involves the oxidation of bromine. As the reaction proceeds, the bromide part of the compound is released from the compound and loses some electrons. These electrons have been transferred to the chlorine atom. So the chlorine atom has been reduced.

While it is not possible to see the movement of electrons, the chemical rules that govern how an element or compound behaves state that electrons are transferred or, in some cases, shared.

USES FOR DISPLACEMENT REACTIONS

Chlorine is a useful substance for chemists because it is good at removing electrons from other substances. This is the principle that is applied in the process of bleaching paper. Chlorine is so effective at taking electrons from other substances that it can change their color. The colors of substances are determined by the number and position of their electrons. If electrons have been removed, the color may change. The same principle also applies when bleach is spilled on clothes.

◀ This wood pulp has been bleached with chlorine. It will be poured into flat sheets and dried to make paper.

TEST YOURSELF

▶ For each of the following reactions, state which substance has been reduced and which has been oxidized. Give reasons for your choices.

(1) $CuO + H_2 \longrightarrow Cu + H_2O$

(2) $2Mg + O_2 \longrightarrow 2MgO$

(3) $2Mg + SO_2 \longrightarrow 2MgO + S$

INVESTIGATE

▶ Use the library or the Internet to find out about the displacement reactions of a group of elements called the halogens. How does reactivity vary in this group, and how does this result in displacement reactions?

Reversible reactions

Some reactions are described as reversible because they can proceed in both forward and reverse directions. The production of ammonia involves a reversible reaction. Ammonia is an extremely important chemical that is used as a refrigerant or a fertilizer. It can also be used in water purification, cleaning products, and explosives.

A SIMPLE REVERSIBLE REACTION

Blue copper sulfate crystals are heated using a Bunsen burner to produce grayish white crystals of anhydrous copper sulfate. The chemical equation for this change is:

$$CuSO_4 \cdot 5H_2O_{(s)} \xrightarrow{heat} CuSO_{4(s)} + 5H_2O_{(g)}$$

Blue copper sulfate crystals contain water molecules. When the crystals are heated the water evaporates, leaving behind anhydrous copper sulfate. This reaction can be reversed by adding water to the grayish white crystals. The crystals will again appear blue and the following

▼ This anhydrous copper sulfate is being hydrated by the addition of water. If the hydrated copper sulfate is heated, this reaction will be reversed.

chemical change has occurred:

$$CuSO_{4(s)} + 5H_2O_{(l)} \longrightarrow CuSO_4 \cdot 5H_2O_{(s)}$$

EQUILIBRIUM

Some reversible reactions proceed in two directions at the same time. Reactions that occur forward and backward at the same time are said to be in equilibrium. Such reactions are only possible if the products and reactants are contained; they must be in close proximity to continue their reactions.

AMMONIA PRODUCTION

Nitrogen and hydrogen gases react to form a gas called ammonia (NH_3).

$$N_{2(g)} + 3H_{2(g)} \longrightarrow 2NH_{3(g)}$$

As soon as the ammonia is formed, it begins to decompose back to nitrogen and hydrogen gases.

$$2NH_{3(g)} \longrightarrow N_{2(g)} + 3H_{2(g)}$$

Because the two reactions occur at exactly the same time, we use the symbol $\rightleftharpoons$. This symbol indicates the reaction is spontaneously reversible. We rewrite the chemical equation as:

$$N_{2(g)} + 3H_{2(g)} \rightleftharpoons 2NH_{3(g)}$$

THE PROBLEM WITH REVERSIBLE REACTIONS

In nonreversible reactions, two or more reactants combine and the reaction ends when all of the reactants are converted into products. Once it is complete, the products do not change back. The reactants fully convert to products.

In a reversible reaction, there is no definite end, and a complete chemical reaction is never achieved. When chemists produce industrial chemicals, they try to obtain as much product as possible before it is converted back to reactants. Producing ammonia in a laboratory involves small amounts of gases, and the cost is minimal.

When ammonia production takes place on a larger scale, as it does in the Haber process, not all of the reactants are converted into ammonia, and a lot of waste is produced. In fact, on average, ammonia producers only convert approximately 15 percent of the reactants. Industrial chemists have improved the efficiency of ammonia production by taking the following steps:

▶ Carry out the reaction at a temperature of approximately 842°F (450°C).
▶ Carry out the reaction under pressure.
▶ Carry out the reaction in the presence of a catalyst.
▶ Pump unreacted gases through the reactor so they are not wasted.

THE HABER PROCESS

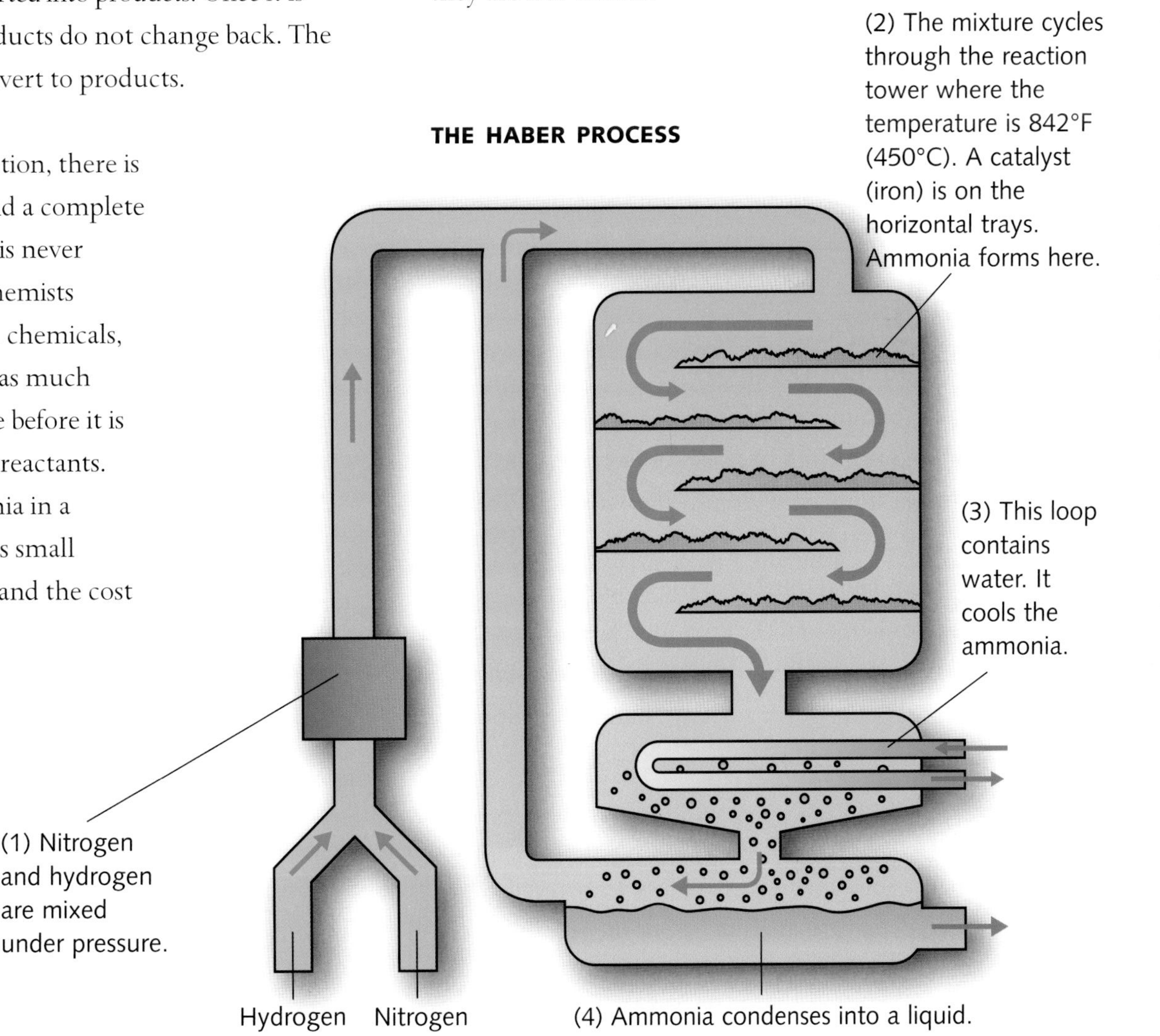

(2) The mixture cycles through the reaction tower where the temperature is 842°F (450°C). A catalyst (iron) is on the horizontal trays. Ammonia forms here.

(3) This loop contains water. It cools the ammonia.

(1) Nitrogen and hydrogen are mixed under pressure.

Hydrogen   Nitrogen

(4) Ammonia condenses into a liquid.

Electrolysis

Electrolysis is a process that uses electricity to separate bonded elements from their compounds. Before a substance can be electrolyzed, it must either be melted or dissolved in a liquid that will conduct electricity. Electrolysis is used in the manufacture of sodium, aspirin, hydrogen, and chlorine. It is also used in submarines to produce oxygen from seawater so that the crew can breathe.

ELECTROLYSIS OF MOLTEN COMPOUNDS

Electricity is the movement of electrons. Electrons are found in the atoms of all substances. Some of a metal's electrons are free to move around the whole substance. As they do so, they carry their negative electric charge with them. Because these electrons are not attached to any particular atom, they are said to be **delocalized**. Solids, such as salt, do not have free electrons, so they cannot conduct electricity in this physical state.

SALT (SODIUM CHLORIDE)

When sodium and chlorine combine chemically, each sodium atom transfers an electron to a chlorine atom. Each sodium atom now has one less electron, and each chlorine atom has one more electron. The result is that the sodium ions have a positive charge and the chlorine ions have a negative charge. Ions with opposite charges attract each other and form a strong chemical compound called sodium chloride.

WHAT HAPPENS TO SALT WHEN IT MELTS?

When a solid melts, its structure becomes less ordered, and the ions are free to move away from each other and carry an electrical charge. The molten salt will eventually decompose into

STRUCTURE OF SOLID SODIUM CHLORIDE

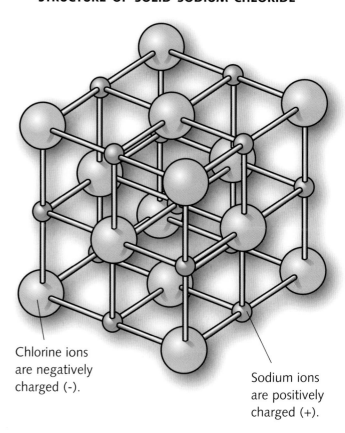

Chlorine ions are negatively charged (-).

Sodium ions are positively charged (+).

▲ Solid sodium chloride forms a large ionic lattice arrangement. The lines represent strong bonds between the atoms.

sodium and chlorine atoms. This also happens when electricity passes through the salt. The decomposition of a substance using electricity is called electrolysis.

MOLTEN LEAD BROMIDE

Lead bromide is electrolyzed to obtain pure lead and pure bromine. When lead bromide is melted, the lead and bromide ions break apart and are free to move. The molten lead bromide is called the electrolyte because it is the chemical that conducts electricity.

WHAT HAPPENS WHEN THE ELECTRICITY IS TURNED ON?

The atoms in the lead bromide decompose into positively charged lead ions and negatively charged bromide ions. Positive ions are called **cations**. Negative ions are called **anions**. The cations (lead ions) are attracted to the electrode that is connected to the negative terminal of the electricity supply, the **cathode**. The anions (bromide ions) are attracted to the positive electrode, the **anode**.

The lead and bromide are now separated from each other. Bromide ions lose their electrons and form bromine fumes at the anode. The electrons that have been released go to the cathode. Here they combine with the lead ions to form lead metal. The lead is in liquid form because the temperature is high.

The overall reaction is:

Lead bromide $\longrightarrow$ Lead + Bromine

$PbBr_{2(l)} \longrightarrow Pb_{(l)} + Br_{2(g)}$

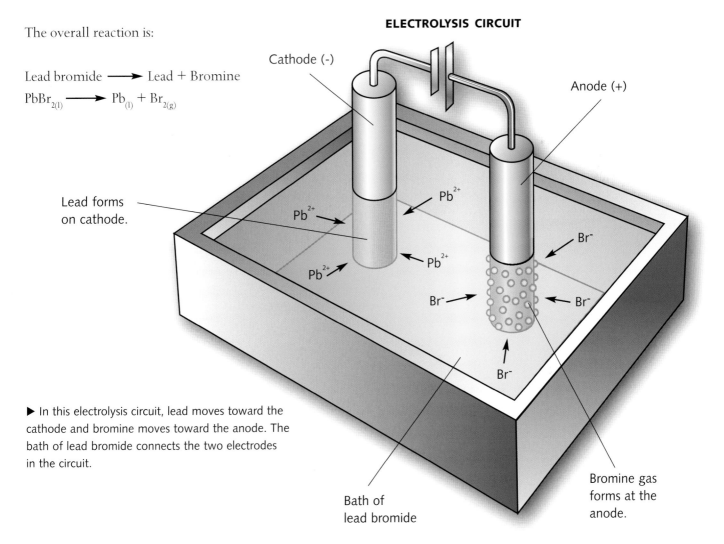

ELECTROLYSIS CIRCUIT

Cathode (-)

Anode (+)

Lead forms on cathode.

Pb^{2+}

Pb^{2+}

Pb^{2+}

Pb^{2+}

Br^-

Br^-

Br^-

Br^-

▶ In this electrolysis circuit, lead moves toward the cathode and bromine moves toward the anode. The bath of lead bromide connects the two electrodes in the circuit.

Bath of lead bromide

Bromine gas forms at the anode.

ELECTROLYSIS OF WATER

Water can be separated into hydrogen and oxygen by electrolysis. Pure water contains some positive hydrogen ions (H^+) and negative hydroxide ions (OH^-). Water is a weak conductor of electricity. When the electricity is turned on, negative hydroxide ions are attracted to the anode. At the anode, hydroxide ions lose their extra electrons and form oxygen gas. The displaced electrons are transferred to the cathode where they react with the positive hydrogen ions. Hydrogen gas is formed. The complete equation is:

Water $\longrightarrow$ Hydrogen + Oxygen
$$2H_2O_{(l)} \longrightarrow 2H_{2(g)} + O_{2(g)}$$

ELECTROLYSIS OF SOLUTIONS

When sodium chloride is dissolved in water, the Na^+ and Cl^- ions mix with the hydrogen ions (H^+) and hydroxide ions (OH^-) from the water.

WHAT HAPPENS AT THE CATHODE?

The H^+ ions and the Na^+ ions are attracted to the cathode. Hydrogen readily receives electrons that come from the anode—more readily than other positive ions. First, the hydrogen ion becomes a hydrogen atom, and then two atoms join to form a hydrogen molecule. Sodium is not produced because hydrogen so quickly accepts the electrons.

TIME TRAVEL: INTO THE FUTURE

Oil is a limited, nonrenewable resource, and oil supplies will be exhausted in your lifetime. Scientists are researching new ways to power cars. Fuel cells are one possible energy source; they were first used during the 1960s in the Gemini-Apollo space programs to generate electrical power and produce drinking water. Fuel cells convert hydrogen and oxygen into water. This produces electricity that can be used to power other devices. Fuel cells are of particular interest to chemists who are concerned with environmental issues because:

▶ They have no moving parts, which means that they do not require a lot of energy.

▶ They do not emit any gases that can harm the earth's atmosphere. The only waste product is water.

The reaction that occurs in a hydrogen fuel cell is:

Hydrogen + Oxygen $\longrightarrow$ Water
$$2H_{2(g)} + O_{2(g)} \longrightarrow 2H_2O_{(g)}$$

◀ This hydrogen fuel cell bus is being tested in Stuttgart, Germany.

WHAT HAPPENS AT THE ANODE?

Cl⁻ and OH⁻ ions are attracted to the anode, but only chlorine gas is released. This is because chloride ions more easily give up their electrons than hydroxide ions. The products formed during the electrolysis of aqueous solutions depends on the ability of the ions to be reduced or oxidized.

USES FOR ELECTROLYSIS

In a process called anodizing, a piece of metal, such as aluminum, is made into the anode. The electrolyte contains a dye material. When the electricity is turned on, the aluminum anode absorbs the color of the dye. This produces colored metals, which are commonly used in metal products, such as bicycle frames. Anodizing is preferable to painting because the dye does not wear off.

The same principle applies to covering one metal with another; this is called electroplating. Electroplating covers steel parts of car bumpers and plates silverware with silver.

ELECTROPLATING

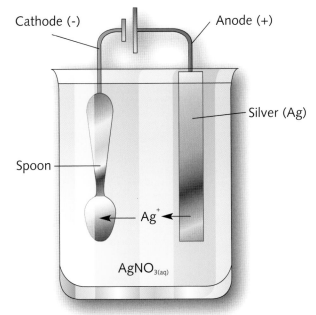

▲ A strip of silver is used to plate a spoon.

▼ Bicycle frames are anodized to give them color.

Analyzing chemical reactions

Chemists often analyze chemicals to determine their components. For example, forensic scientists analyze unknown substances found at crime scenes to determine what they are and whether they are related to the crime or the criminal. Chemists test the products of a chemical reaction in order to decide what reaction has occurred. The analysis of chemicals is methodical and involves many different tests.

TESTING FOR METAL IONS

If a chemist suspects that a substance contains metal ions (cations) there are two types of tests to conduct—a flame test and a precipitation test.

▶ FLAME TEST

When some metals are placed into a blue flame, they produce distinctive colors. The chemist observes the color and identifies the metal ion involved.

HOW TO CARRY OUT A FLAME TEST:

(1) Dip a nichrome wire into concentrated hydrochloric acid and place it into a blue flame. Nichrome is a mixture of the metals nickel and chromium and it will not change color if placed into a blue flame.

(2) If you can see a color, repeat the first step until no color appears. The wire is now clean.

(3) Dip the nichrome wire into concentrated hydrochloric acid, and then into the sample to be tested. The sample should stick to the damp wire.

(4) Hold the metal sample in the flame and observe the color change.

▲ A positive lithium flame test

Characteristic colors are produced for some metal ions, as shown in the following table:

Metal ion	Formula of ion	Color
Lithium	Li^+	Scarlet red
Sodium	Na^+	Yellow
Potassium	K^+	Lilac
Barium	Ba^{2+}	Apple green
Copper	Cu^{2+}	Green
Calcium	Ca^{2+}	Brick red

▶ PRECIPITATION TEST

When two solutions are mixed and an insoluble solid forms, precipitation has occurred. Certain metal ions produce distinct colored precipitates.

HOW TO CARRY OUT A PRECIPITATION TEST:

(1) Mix the test substance with sodium hydroxide solution.

(2) Observe the color of the precipitate. Characteristic colors are produced for some metal ions, as shown in the following table:

Metal ion	Formula of ion	Color of precipitate with sodium hydroxide
Iron (II)	Fe^{2+}	Green
Iron (III)	Fe^{3+}	Rusty brown
Copper (II)	Cu^{2+}	Blue
Nickel (II)	Ni^{2+}	Green
Cobalt (II)	Co^{2+}	Pink

◀ A positive precipitation test for copper.

Notice how a metal can form more than one type of metal ion. Iron can form two types of ions, depending on how many electrons it has lost. Iron (II) has lost two electrons, whereas iron (III) has lost three electrons.

TESTING FOR AMMONIUM IONS

Ammonium ions have the formula NH_4^+. They are **complex ions** because they are composed of more than one type of element—in this case, nitrogen and hydrogen. Ammonium ions are not metallic and will not test positive with a flame or precipitation test. When chemists suspect that a solution contains ammonium ions, the solution is heated and the gas that is emitted is tested with damp red litmus paper. The paper will turn blue as ammonia gas is given off. Ammonia gas also has a distinctive, pungent smell.

TESTING FOR NONMETAL IONS

In addition to testing for the metal ions contained in a substance, chemists also conduct tests to identify the nonmetallic parts, or anions, in a substance. These anions have their own chemical tests (see the following information on pages 34 and 35).

TEST YOURSELF

▶ A forensic chemist was asked to identify three solutions using precipitation tests. The chemist knew that one of the solutions contained copper (II) ions, another iron (II), and the last iron (III) ions. What would the chemist do to identify the solutions and what results could be observed?

TESTING FOR HALIDES

Halides are nonmetal ions from a group on the periodic table called the halogens. They include chloride (Cl^-), bromide (Br^-), and iodide (I^-) ions. Silver nitrate is used to test for halide ions, as follows:

(1) Dilute nitric acid is added to the test substance to remove any impurities.
(2) Silver nitrate solution is introduced; if a halide ion is present, a distinct, colored precipitate forms.

Anion	Formula of anion	Color of precipitate	Name of precipitate	Formula of precipitate
Chloride	Cl^-	White	Silver chloride	AgCl
Bromide	Br^-	Cream	Silver bromide	AgBr
Iodide	I^-	Pale yellow	Silver iodide	AgI

The reactions that occur are:

Silver nitrate + Chloride ⟶ Silver chloride + Nitrate

Silver nitrate + Bromide ⟶ Silver bromide + Nitrate

Silver nitrate + Iodide ⟶ Silver iodide + Nitrate

Each of these three precipitates is sensitive to light. If they are left in bright sunshine, they turn gray.

TESTING FOR COMPLEX ANIONS

Complex anions contain more than one type of element. We can tell from their formula that they are complex because they have more than one capital letter. For example, a sulfate anion has the formula SO_4^{2-}. This tells us it has a negative charge, as we can see from the $^{2-}$, and consists of the elements sulfur (S) and oxygen (O). Other complex anions include hydroxide (OH^-) and carbonate (CO_3^{2-}).

TESTING FOR SULFATES

Chemists use a precipitation reaction to test for sulfates.

▲ From left to right, these test tubes contain silver chloride, silver bromide, and silver iodide.

(1) Dilute nitric acid is reacted with a substance.
(2) A barium chloride solution is added.

A white precipitate indicates the presence of a sulfate. This precipitate is called barium sulfate and is not light-sensitive. The reaction that has occurred is:

Barium chloride
+
Sodium sulfate
⟶
Barium sulfate
+
Sodium chloride

TESTING FOR CARBONATES

All carbonates react with acids to produce carbon dioxide gas.
Here is one example:

Calcium carbonate + Hydrochloric acid $\longrightarrow$ Calcium chloride + Water + Carbon dioxide

$$CaCO_{3(s)} + 2HCl_{(aq)} \longrightarrow CaCl_{2(aq)} + H_2O_{(l)} + CO_{2(g)}$$

When carbon dioxide is bubbled through calcium hydroxide
solution, the solution turns milky.

▲ This forensic scientist is collecting evidence from a window.
By chemically analyzing trace (very small) chemicals she
may be able to find out what smashed the window.

TEST YOURSELF

▶ Which elements, and how many atoms of each
element, are present in the following complex anions?

Hydroxide (OH⁻), Carbonate ($CO_3{}^{2-}$)

▶ Flame tests were carried out on three unknown
compounds: A, B, and C. The results are given below.

Compound	Color of flame
A	Lilac
B	Green
C	Yellow

Dilute nitric acid was then added to each
compound. Next, a solution of silver nitrate was
added. The results are given below.

Compound	Color of precipitate
A	Yellow
B	White
C	Cream

What are A, B, and C?

TIME TRAVEL: INTO THE FUTURE

In the future, chemical reactions may be used
to create life. At the Los Alamos National
Laboratory in the U.S., a scientist named Steen
Rasmussen has been awarded a $5 million
grant to pursue this research. He and his team
hope to build a "protocell," which will be a life-
form smaller than a bacterium. This unique
"creature" will not resemble any life-form that
we currently know.

The ingredients for making the life-form are a chemical
called PNA, a chemical called pinacol, and "precursor
molecules" that act as food for the protocell. In an
incredible series of chemical reactions, the scientists
hope that the life-form will produce energy, grow, and
reproduce. If they succeed, the researchers will be
crossing the barrier between natural and human
powers. Their work could provide new information
about the origins of life on the earth.

Biological reactions

For thousands of years, humans have used organisms to create useful substances through chemical reactions. A type of fungus called yeast is one of these organisms. Yeast contains chemicals called enzymes. Enzymes are biological catalysts that accelerate the conversion of reactants into products. The yeast enzyme is called zymase. It aids the conversion of glucose into ethanol, which has many uses as fuels, drinks, and as a solvent.

RESPIRATION

Respiration is the process that living organisms use to convert energy from their food into energy that they can use for living processes. When yeast and sugar are combined and the reactants are kept warm, the following respiration reaction occurs.

$$\text{Glucose} \xrightarrow{\text{yeast}} \text{Ethanol} + \text{Carbon dioxide}$$

This process releases energy and proceeds without oxygen. We call this **anaerobic respiration**.

CONDITIONS FOR ETHANOL PRODUCTION

When ethanol is produced on an industrial scale, the chemical reaction occurs in large containers called fermenters. The organisms inside the fermenters require the following conditions to ensure an efficient reaction.

▶ **FOOD** – Yeast organisms need glucose as their food source. Different types of sugars produce different types of alcoholic drinks. For example, wine is the fermentation product of grapes.

▼ This vat contains red wine. As the wine ferments, it forms froth—bubbles of carbon dioxide.

▶ Optimum temperature conditions —

Enzymes tend to function most efficiently at about 98.6°F (37°C). If the temperature is too high, the enzyme no longer functions. This process is called **denaturing**. If the temperature is too low, then fermentation will be very slow.

▶ Limited amount of oxygen —

Yeast converts glucose to ethanol when there is a limited supply of oxygen. The air in the fermenter is mixed with extra carbon dioxide to limit the concentration of oxygen. The fermenter is sealed. Carbon dioxide produced during the reaction is allowed to escape through valves.

▶ Mixing —

Industrial fermenters are fitted with stirrers. Constant stirring ensures that as much of the sugar as possible is converted into alcohol.

However, not all of the sugar is converted to alcohol; as the concentration of alcohol builds, it destroys the yeast. The yeast becomes poisoned by the alcohol.

Uses of alcohol

Alcoholic drinks are a part of many societies, but this is not the only use for alcohol. Ethanol is also an important solvent and is used in the manufacture of perfumes and aftershave. The fragrance is dissolved into ethanol, and when this is sprayed onto the skin, the ethanol quickly evaporates and leaves the scent behind.

In Brazil, sugarcane is grown on a vast scale. It is fermented to produce ethanol that is mixed with gasoline and used to fuel cars. The mixture is called "gasohol."

Spoilage

Chemical reactions occur when bacteria come into contact with food. Bacteria feed on food and spoil it. Eating food with bacteria on it can cause sickness, such as vomiting and diarrhea. No bacteria are allowed access to the alcohol in a fermenter.

Time travel: Discoveries of the past

During the late eighteenth and early nineteenth centuries, Napoleon's armies were widely dispersed between Russia and Spain. They suffered from a lack of fresh food because their supply-line was long and vulnerable to attack. In 1795, Napoleon offered a prize of 12,000 francs to anyone who could solve the food supply problem. The prize was awarded in 1812 to Nicolas Appert, a French confectioner. Appert invented a method of sterilization. He cooked food and placed it into an airtight glass jar, while it was still hot. This was the birth of the modern tin can.

How to speed up a reaction

Chemical reactions can happen instantly, or they can occur over many years or decades. For example, when sodium reacts with water, the change is almost instantaneous. On the other hand, copper roofing materials gradually oxidize and turn green over many years. Often, chemists speed up chemical reactions, particularly when they need to manufacture large quantities of chemicals in short periods of time.

REACTION RATES

The rate of a chemical reaction is a measure of the amount of reactant that turns into product in a unit of time. The rate is important to a manufacturer because they need to meet the demand for their product and ensure that they make a reasonable profit. The rate of a chemical reaction is not written as part of a balanced chemical equation. The only way to find out the rate of reaction is to conduct the experiments.

It is also important to realize that the reaction rate changes during a single reaction. At the beginning of a reaction, a lot of reactant is present. As the reaction proceeds, some reactant is converted into product. The product stops reacting, and the reaction rate slows down. Eventually, when all the reactant has been converted into product, the reaction is complete. This can be displayed graphically.

RATE OF REACTION

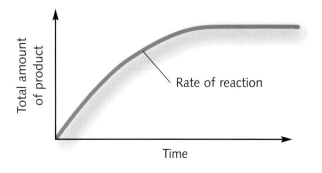

Total amount of product

Rate of reaction

Time

TEMPERATURE AND REACTION RATES

Temperature has a significant effect on reaction rates. Bunsen burners heat chemicals in a laboratory to accelerate a reaction. In our homes, we place food in the refrigerator to slow down the reactions that spoil it. In fact, scientists have discovered that if the temperature of some reactions is increased by 50°F (10°C), the rate of the reaction is almost doubled.

▲ Increasing the temperature of a reaction usually increases the reaction rate.

COLLISION THEORY—TEMPERATURE

During a chemical reaction, the reacting particles must collide with each other in order for the reaction to occur. However, when particles collide, they must meet the following conditions:

▶ Reactants must collide with energy that is sufficient to break the chemical bonds.

▶ Reactants must collide at an angle that ensures all of their energy is spent on the reaction.

AN UNSUCCESSFUL COLLISION

A SUCCESSFUL COLLISION

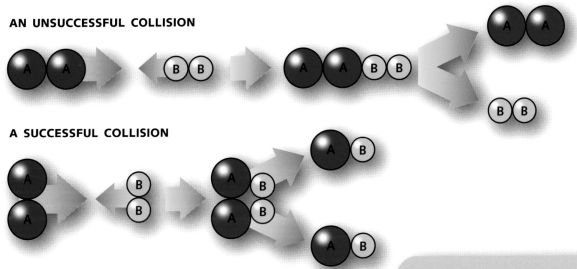

▲ The direction and angle with which the reactants collide affects the success of the chemical reaction.

As the temperature of a reaction increases, the following happens:

(1) The reacting particles gain energy and move faster.
(2) The reacting particles collide more frequently.
(3) The reacting particles have more energy and the collisions are more likely to lead to a reaction.

RATE OF REACTION

Higher temperature

Lower temperature

Total amount of product

Time

TEST YOURSELF

▶ Two students reacted magnesium metal with sulfuric acid and measured the rate of reaction. The reaction was carried out at room temperature—68°F (20°C). The results are shown below.

Volume of gas (cm³)	Time (mins)
0	0
5	1
10	2
15	3
17	4
17	5

Plot this data on a graph. Make sure you have labeled the axes correctly. Draw lines on the graph to illustrate the reaction results if it were carried out at 50°F (10°C) and 104°F (40°C).

▶ Use the idea of collision theory to explain why milk is kept in the refrigerator.

CONCENTRATION AND REACTION RATES

When you make a cup of hot chocolate, the instructions on the container tell you how many spoonfuls of chocolate powder to add. If you want a drink that is richer, you add more chocolate powder. This would give you a more concentrated drink.

The concentration of a solution is the number of particles that have been dissolved into a given volume of water or other solvent. A high concentration has a lot of particles dissolved into it, whereas a low concentration has fewer particles. If concentrations of chemical solutions in a reaction are increased, the rate of reaction also increases. However, the final amount of end product does not increase.

REACTING MAGNESIUM WITH HYDROCHLORIC ACID

The chemical equation for this reaction is:

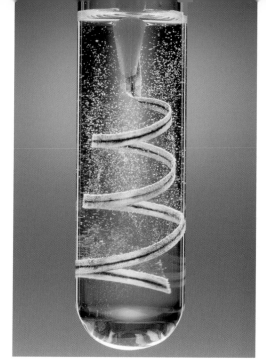

▲ This magnesium ribbon is reacting with hydrochloric acid. The more concentrated the acid, the faster the reaction will proceed.

Magnesium + Hydrochloric acid $\longrightarrow$ Magnesium chloride + Hydrogen

$$Mg_{(s)} + 2HCl_{(aq)} \longrightarrow MgCl_{2(aq)} + H_{2(g)}$$

It is possible to accelerate this reaction by increasing the concentration of the hydrochloric acid, while all other conditions are unchanged.

A chemist carried out this reaction with two samples of hydrochloric acid. Sample I was twice as concentrated as sample II. The chemist collected the hydrogen gas produced with each reaction. The graph below shows the results of these reactions. Comparing the two lines on the graph, the following becomes obvious:

(1) The line for sample I is much steeper than the line for sample II. This means that sample I produced more hydrogen gas than sample II in the same amount of time.

(2) Both lines stop at the same volume of gas, because the quantity of reactant was the same in each reaction. Sample II reacted more slowly because it had fewer reacting particles in the same volume.

REACTING MAGNESIUM WITH HYDROCHLORIC ACID

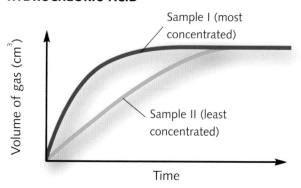

COLLISION THEORY—CONCENTRATION

This theory states that reactants are only converted to products if the reacting particles collide with the required amount of energy, and at an effective angle. If the concentration of a solution is increased, the reaction is faster as there are more reacting particles in a given volume. This means there will be more collisions in a given period of time, and more collisions are likely to lead to a successful reaction and product.

CONCENTRATION AND GASES

The same principle applies to gases under pressure. If a gas is compressed, the same number of gas particles are in a smaller space. The concentration has effectively increased. Therefore, reaction rates between gases are increased by higher pressures.

COMPRESSED GAS

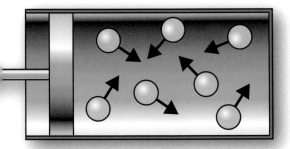 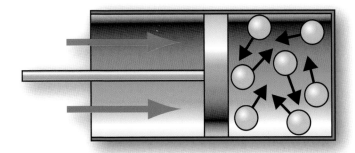

▲ Compare the two syringes. The gas in the syringe on the right has been compressed. The particles have been squeezed into a smaller place. This has increased the concentration.

CATALYSTS AND REACTION RATES

A catalyst increases the reaction rate. Because a catalyst is not chemically changed by the reaction, it can be used over and over again. For example, hydrogen peroxide decomposes very slowly at room temperature. The chemical equation is:

Hydrogen peroxide $\longrightarrow$ Water + Oxygen

$$2H_2O_{2(aq)} \longrightarrow 2H_2O_{(l)} + O_{2(g)}$$

If a bottle of hydrogen peroxide is left open for several years, all that would be left is water—if it has not all evaporated. This reaction can be accelerated by adding a catalyst called manganese (IV) oxide.

Manganese (IV) oxide is a transition metal. Compounds that contain transition metals make excellent catalysts. In the reaction with hydrogen peroxide, the manganese (IV) oxide immediately makes the hydrogen peroxide froth up as oxygen gas is produced. At the end of the reaction, the manganese (IV) oxide remains chemically unchanged and can be seen as a fine black powder in the reacting container.

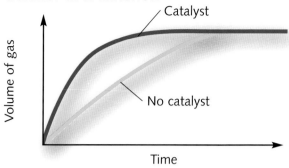

CATALYST IN A REACTION

▲ This graph shows the effect of adding a catalyst to hydrogen peroxide. The catalyst speeds up the reaction.

HOW DO CATALYSTS WORK?

Catalysts provide an alternate route by which a reaction can occur. This is like a person who needs to reach the other side of a mountain. A lot of energy could be spent going over the top of the mountain, or less energy could be expended going through a tunnel that is bored through the middle of the mountain. The result is the same, but the activation energy required is much lower one way than the other. When catalysts are added to reactions, more reactants have the necessary energy and the reactions are much faster.

▶ Hydrogen peroxide in the beaker is decomposing. Manganese (IV) oxide is the catalyst. It speeds up the reaction.

DIFFERENT TYPES OF CATALYST

▶ Catalysts exist in any physical state. If they are in the same physical state as the reactants, they are called **homogeneous**. Enzymes speed up the many reactions in our bodies. They work in an aqueous solution, along with the reactants that they act upon.

▶ Catalysts that are in a different physical state than the reactants are called **heterogeneous**. The reaction involving manganese (IV) oxide and hydrogen peroxide is heterogeneous. Manganese (IV) oxide is a solid and hydrogen peroxide is in solution.

CATALYTIC CONVERTERS

All new cars are fitted with catalytic converters. They convert harmful exhaust fumes into less harmful products. Catalytic converters are made from the precious metals platinum and rhodium. This is why it is expensive to buy a catalytic converter. Cheaper metals are not used because they are easily degraded by the exhaust fumes.

Despite the fact that platinum and rhodium are expensive, their use is economically viable; if treated properly, they will never need to be replaced and they will never be used up. In addition, both metals can be extracted from the catalytic converters of scrapped cars and recycled.

WHAT HAPPENS IN A CATALYTIC CONVERTER?

(1) Carbon monoxide is converted into carbon dioxide.

(2) Partially burned gasoline is converted into carbon dioxide and water.

(3) Nitrogen oxides are converted into harmless nitrogen and oxygen.

(4) These safer gases are then emitted from the exhaust pipe.

▲ These cars are on their way to be recycled. Platinum and rhodium are retrieved from scrap cars and made into new catalytic converters.

TEST YOURSELF

▶ Hydrogen peroxide can also be broken down by a biological enzyme called catalase. Design an experiment to test which is better at breaking down hydrogen peroxide—catalase or manganese (IV) oxide.

SURFACE AREA AND REACTION RATES

In chemical reactions, larger pieces of reactant react more slowly than smaller pieces. When a solid is involved in a chemical reaction, the reaction takes place on the surface of the solid. The more surface that is exposed, the greater the rate of reaction.

Study the cube on this page. When the cube is cut in half, the exposed surface area increases, but the total volume remains the same. Therefore, when the cube is cut in half, there is a higher surface area-to-volume ratio than if there is just one large cube. The higher the surface area-to-volume ratio, the faster the rate of reaction.

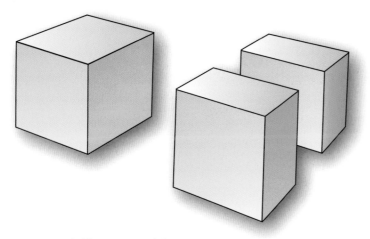

▲ The more a solid is divided, the greater the surface area-to-volume ratio.

COLLISION THEORY—SURFACE AREA

When the number of successful particle collisions increases, the rate of chemical reaction will increase. Collisions take place on the surface of a solid. Therefore, a greater surface area-to-volume ratio will result in a faster reaction rate.

LIMESTONE AND ACID

The following chemical equation shows the reaction between calcium carbonate (limestone) and hydrochloric acid:

Calcium carbonate + Hydrochloric acid $\longrightarrow$ Calcium chloride + Water + Carbon dioxide

$$CaCO_{3(s)} + 2HCl_{(aq)} \longrightarrow CaCl_{2(aq)} + H_2O_{(l)} + CO_{2(g)}$$

The graph below shows what happens when different-sized pieces of limestone, all of which have the same mass, react with the same concentration of hydrochloric acid.

REACTION RATE AND SURFACE AREA OF REACTANTS

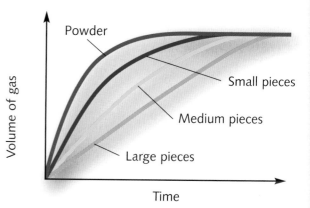

▲ The smaller the particle size, the greater the surface area-to-volume ratio, and the more rapidly the reaction proceeds. The amount of final product remains the same.

▲ Both beakers contain hydrochloric acid and calcium carbonate—as a powder on the left and as a single piece on the right. The reaction rate is faster when the surface area is greater.

The reaction occurs on the surface of the limestone pieces. When the limestone is in powder form, the maximum surface area is exposed and the reaction is almost instantaneous. A large amount of gas is released very quickly. If large pieces of limestone are used, a much smaller area is available for reaction. As the outer layer of limestone reacts, a fresh layer is exposed and is now available for reaction. The reaction rate appears to be more gradual, and it may appear as if not much is happening. Because all of these reactions involve the same mass of limestone, the final amount of product is the same.

MEASURING MASS TO DETERMINE REACTION RATE

The reaction of limestone and hydrochloric acid can also be measured by calculating the loss in mass. Gas produced by the reaction could escape the container holding the reactants and the acid may spray, so a cotton wool plug is necessary for safety and to prevent an unnecessary loss in mass. As carbon dioxide escapes, the mass of the contents of the container decreases. Mass is recorded at regular intervals, demonstrating a gradual loss in mass. When the mass measurements are plotted on a graph, the reaction rate can be determined.

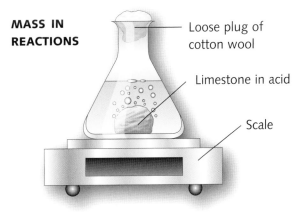

MASS IN REACTIONS

— Loose plug of cotton wool

— Limestone in acid

— Scale

▲ The mass will decrease as the reaction between the limestone and the acid proceeds.

▲ You can see the bubbles of carbon dioxide escaping from the outside layer of this piece of limestone.

INVESTIGATE

▶ **Ask an adult for permission before you do this experiment.**
Investigate the effect of an increased surface area-to-volume ratio by melting cheese on toast. Weigh slices of cheese and place them on a slice of bread. Grate the same mass of cheese onto another slice of bread. Place both slices under the oven broiler. Which slice of bread and cheese cooks first? Why?

Glossary

ANAEROBIC RESPIRATION – Respiration in the absence of oxygen.

ANION – A negatively charged atom.

ANODE – A positive electrode.

BOIL – To heat a liquid to its boiling point. At this temperature, it will undergo a physical change from a liquid to a gas.

CATALYST – A substance used to speed up a chemical reaction without being used up or changed.

CATHODE – A negative electrode.

CATION – A positively charged atom.

CHEMICAL BONDS – The forces that hold two or more atoms together in a substance.

COMBUSTION – The reaction between a fuel and oxygen that produces heat or light. This is also called burning.

COMPLEX ION – An ion composed of more than one type of atom.

CONDENSE – A physical change from a gas to a liquid.

CORRODE – The chemical decomposition of a substance. The rusting of iron is a type of corrosion.

DELOCALIZED – A delocalized electron is free to move, rather than being bound to a particular atom.

DENATURE – To permanently alter the shape of a protein, such as an enzyme.

DIFFUSION – The movement of a substance from an area of high concentration to an area of low concentration.

DISPLACEMENT – When one substance is removed from a compound by another, more reactive, substance.

ELECTROLYSIS – This process involves passing electricity through a compound to split it into its elements.

ENDOTHERMIC – Reactions in which overall heat energy is absorbed from the surroundings. Endothermic reactions feel cold.

ENZYME – A biological catalyst.

EVAPORATION – A physical change from a liquid to a gas.

EXOTHERMIC – Reactions in which overall heat energy is released.

FREEZE – A physical change from a liquid to a solid.

HETEROGENEOUS – Two substances in different physical states.

HOMOGENEOUS – Two substances in the same physical state.

IMMISCIBLE – Substances that will not mix.

MELT – A physical change from a solid to a liquid.

MISCIBLE – Substances that will mix together.

OXIDIZER – A chemical that contains a lot of oxygen and can oxidize another chemical.

PARTICULATES – Fine particles in air or water.

PRECIPITATE – An insoluble solid formed when two solutions react.

PRODUCTS – Chemicals resulting from a chemical reaction.

REACTANTS – Initial materials for a chemical reaction.

REDOX – Reduction and oxidation reactions occurring together.

SUSPENSION – A fine solid distributed in a liquid.

Useful Web sites:

http://www.chem4kids.com
http://www.howstuffworks.com
http://www.chemtutor.com
http://www.sciencenewsforkids.org
http://www.newscientist.com

Page 23: Test yourself
Combustion takes place in central heating systems, in a car's engine, when a match is lit, and in coal-burning power stations that produce electricity.

Page 25: Test yourself
(1) CuO is reduced because it has lost oxygen; H_2 is oxidized because it has gained oxygen.
(2) Mg has been oxidized because it has gained oxygen; therefore, O must be reduced.
(3) Mg is oxidized because it has gained oxygen; SO_2 has lost oxygen, so it is reduced.

Page 25: Investigate
Halogens at the top of the group are more reactive than those at the bottom. The more reactive halogens can displace less reactive halogens from their compounds.

Page 31: Test yourself
(1) Sodium and chlorine, (2) Copper and iodine

Page 33: Test yourself
Add sodium hydroxide solution. Copper (II) gives a blue precipitate, Iron (II) a green precipitate, and Iron (III) a rust-colored precipitate.

Page 35: Test yourself
Hydroxide = one oxygen and one hydrogen atom. Carbonate = one carbon and three oxygen atoms. A is potassium iodide (KI), B is copper chloride ($CuCl_2$), and C is sodium bromide (NaBr).

Page 39: Test yourself
The graph will be a curve with a shape similar to that on page 38. After five minutes at 50°F (10°C), the line will be more shallow and will not go any higher than about 10 cm³. After five minutes at 104°F (40°C), the line will rise more sharply but will go no higher than 17 cm³. Milk is stored at a low temperature because particles that react and spoil it have less energy at lower temperatures, so the reaction is much slower.

Page 41: Test yourself
(1) Increase the number of students. (2) Partition some of the playground so that the students are in a smaller space.

Page 43: Test yourself
Use the same amount of hydrogen peroxide for testing both catalysts. Use the same mass of each catalyst. Record the amount of froth produced in a given unit of time for each catalyst. The catalyst that produces the most froth is better at decomposing hydrogen peroxide.

Page 45: Investigate
The slice of bread with grated cheese cooks first. It has a greater surface area-to-volume ratio.

Index

Page references in *italics*
represent pictures.

Photo Credits – (abbv: r, right, l, left, t, top, m, middle, b, bottom) **Cover background image** www.istockphoto.com/Jon Helgason **Front cover images** (r) www.istockphoto.com/jallfree (l) www.istockphoto.com/foto pfluegl **Back cover image** (inset) www.istockphoto.com/jallfree **p.1** (t) David Taylor/Science Photo Library (br) NASA (bl) Tek Image/Science Photo Library **p.2** www.istockphoto.com/Justin Allfree **p.3** (b) www.istockphoto.com/Klaas Lingbeek-van Kranen (t) Erich Schrempp/Science Photo Library **p.4** (tr) www.istockphoto.com/Feng Yu (tl) Andrew Lambert Photography/Science Photo Library (br) Martin Bond/Science Photo Library **p.5** Charles D. Winters/Science Photo Library **p.7** Kip Peticolas/Fundamental Photo/Science Photo Library **p.8** Erich Schrempp/Science Photo Library **p.9** NASA **p.10** Andrew Lambert Photography/Science Photo Library **p.11** www.istockphoto.com/Ethan Gibbs **p.12** Robert Brook/Science Photo Library **p.14** (both) Martyn F. Chillmaid/Science Photo Library **p.15** Martyn F. Chillmaid/Science Photo Library **p.19** Charles D. Winters/Science Photo Library **p.20** (t) www.istockphoto.com/Jonathan Klemenz (m) Andrew Lambert Photography/Science Photo Library **p.21** (t) Sheila Terry/Science Photo Library (b) Andrew Lambert Photography/Science Photo Library **p.22** (b) Reuters/Corbis **p.23** www.istockphoto.com/Klaas Lingbeek-van Kranen **p.24** Jacques Jangoux/Science Photo Library **p.25** R. Maisonneuve, Publiphoto Diffusion/Science Photo Library **p.26** Andrew Lambert Photography/Science Photo Library **p.30** Martin Bond/Science Photo Library **p.31** (r) www.istockphoto.com/Justin Allfree **p.32** Andrew Lambert Photography/Science Photo Library **p.33** Andrew Lambert Photography/Science Photo Library **p.34** Andrew Lambert Photography/Science Photo Library **p.35** Tek Image/Science Photo Library **p.36** Charles O'Rear/Corbis **p.37** Corbis **p.38** Richard Megna/Fundamental Photos/Science Photo Library **p.40** Charles D. Winters/Science Photo Library **p.41** John Walsh/Science Photo Library **p.42** Charles D. Winters/Science Photo Library **p.43** www.istockphoto.com/Feng Yu **p.44** Martyn F. Chillmaid/Science Photo Library **p.45** David Taylor/Science Photo Library